Who Beats the Heat?

Pamela Chanko • Daniel Moreton

Scholastic Inc.
New York • Toronto • London • Auckland • Sydney

Acknowledgments

Science Consultants: Patrick R. Thomas, Ph.D., Bronx Zoo/Wildlife Conservation Park; Glenn Phillips, The New York Botanical Garden

Literacy Specialist: Ada Cordova, District 2, New York City

Design: MKR Design, Inc.

Photo Research: Barbara Scott

Endnotes: Samantha Berger

Endnote Illustrations: Craig Spearing

Photographs: Cover G. C. Kelly/Photo Researchers; p. 1: Tony Craddock/Photo Researchers; p. 2: G. C. Kelly/Photo Researchers; p. 3: Jerry L. Ferrara/ Photo Researchers; p. 4: Stephen Krasemann/Photo Researchers; p. 5: David Northcott/DRK Photo; p. 6: Stephen J. Krasemann/DRK Photo; p. 7: Jim Steinberg/Photo Researchers; p. 8: Tom & Pat Leeson/Photo Researchers; p. 9: G. C. Kelly/Photo Researchers; p. 10: M.C. Chamberlain/ DRK Photo; p. 11: Jerry L. Ferrara/Photo Researchers; p. 12: Stephen J. Krasemann/DRK Photo.

8   9   10   08   03   02   01   00

Who beats the heat in the desert?

Owls beat the heat.

Jackrabbits do, too.

Foxes beat the heat.

Snakes do, too.

Bobcats beat the heat.

Lizards do, too.

Peccaries beat the heat.

Spiders do, too.

Prairie dogs beat the heat.

Tortoises do, too.

Desert animals know how to beat the heat!

Who Beats the Heat?

The desert (page 1), with its extreme temperatures and scarcity of water and food, can be a hard place to live. These are some of the animals that are able to beat the heat and survive in the desert.

The elf owl (page 2), one of the smallest owls in the world, makes its nest in a saguaro cactus. The inside of the cactus is cool and moist, unlike the outside, which is beaten by harsh sunlight. The saguaro is home to other birds as well as rodents such as mice. Birds have a higher body temperature than mammals, which means they can tolerate higher temperatures outside.

Jackrabbits (page 3) in the desert rest by day and become more active when the sun is not so strong. They try to stay in the shade of vegetation like tall grass or in shallow holes, where the soil is cooler.

The kit fox (page 4) is mostly nocturnal, which means it comes out at night. The desert is much cooler at night! During the day, the kit fox stays in its den or opens the burrows of other small animals in order to get food.

The horned viper (page 5) does not actually have horns; its spine forms these hornlike points. The viper beats the heat by burying itself in the sand, where it's much cooler. It wriggles its body, burrowing deeper and deeper until just its nostrils and eyes are emerging from the sand. Just enough to detect danger!

The bobcat (page 6), the most common wildcat in North America, stays out of the daylight as well. It lies by day in a rock cleft, thicket, or other hiding place. In the evening it emerges and hunts, preying on snowshoe hares, cottontails, mice, cave bats, and even porcupines!

The spiny desert lizard (page 7) is active during the day. But it is very cautious and darts into rock crevices or rodent holes when startled. It can often be found climbing trees or walls in order to find insect prey.

The collared peccary (page 8) is active in the early morning and late afternoon to avoid the midday heat. The rest of the time, the peccary can often be found bedding down in a shallow hole in the earth or a cave, where the temperatures are much cooler.

Even the heavy, hairy desert tarantula (page 9) hides by day in abandoned holes or beneath rocks. The male wanders in the dim light after sunset or near dawn looking for a mate.

Prairie dogs (page 10) stay underground in their burrows to avoid the heat and direct sunlight. Temperatures within a burrow don't change nearly as much as they do on the surface. The deeper the burrow, the better the insulation! During the day the burrow stays cool, and at night the burrow stays warm.

The desert tortoise (page 11) comes out during the early and late afternoon to feed on grasses. During the hottest part of the day, it retreats to a shallow burrow, usually dug in the base of an arroyo wall. It has been known to hollow out horizontal holes up to 30 feet in length!

Unlike most cats, the mountain lion (page 12) is active during the day. But it is a very solitary animal and keeps out of sight. It will travel 25 to 50 miles in order to find prey, which includes large and small mammals, birds, and even grasshoppers!

Science
EMERGENT READERS

The hot, dry desert is a tough place to live.
Find out how different animals
adapt to this climate.

ISBN 0-590-63873-4

90000

9 780590 638739